Frozen FROGS and Sleeping SQUIRRELS

Animals in Winter

by Joanne Mattern

BeaLu
BOOKS

ISBN Hardcover: 978-1-962981-00-2
ISBN Paperback: 978-1-962981-03-3

Library of Congress Control Number: 2023950311
Publisher's Cataloging-in-Publication Data is on file with the publisher.

Edited by: Precious McKenzie
Book cover and interior design by Tara Raymo • creativelytara.com

Printed in the United States of America
November 2023
First Edition
2 7 6 5 1

BeaLu Books
Tampa, Florida

www.BeaLuBooks.com

PHOTO CREDITS: Cover: © Alyena_Krylova – Shutterstock, © DNY3D – Adobe Stock, © sandymsj – Shutterstock; Page 4: © Irina Wilhauk – Shutterstock; Page 5: © yhelfman – Shutterstock; Page 6: © isabel kendzior -Shutterstock; Page 7: © Brett Swain – Shutterstock; Page 8: © photos martYmage – Shutterstock; Page 9: © Jane Rix - Shutterstock , © Kjetil Kolbjornsrud – Shutterstock; Page 10: © L-N – Shutterstock; Page 11: © Ondrej Prosicky - Shutterstock , © BMCL – Shutterstock; Page 12: © Scalia Media – Shutterstock; Page 13: © Erni – Shutterstock; Page 14: © Chris Hill – Shutterstock; Page 15: © Altitude Visual – Shutterstock, © StockMediaSeller – Shutterstock; Page 16: © D. Longenbaugh – Shutterstock; Page 17: © Jess Lang – Shutterstock; Page 18: © RukiMedia – Shutterstock; Page 19: © Denizce – Shutterstock; Page 20: © Payman Sedaghat – Shutterstock; Page 21: © Travel Faery – Shutterstock; Page 22: © MarclSchauer – Shutterstock; Page 23: © Agami Photo Agency – Shutterstock; Page 24: © Scott Thornhill – Shutterstock; Page 25: © TangoFoxtrot2018 – Shutterstock; Page 26: © Real PIX – Shutterstock; Page 27: © Wirestock Creators – Shutterstock, © Dark_Side – Shutterstock; Page 28: © The Sturdy Table – Shutterstock; Page 29: © Sean Pavone – Shutterstock; Page 30: © IrinaK – Shutterstock; Page 31: © Sean Fleming – Shutterstock; Page 32: © Dennis W Donohue – Shutterstock; Page 33: © Kirk Geisler – Shutterstock, © Toni Genes – Shutterstock; Page 34: © Paul Roedding – Shutterstock; Page 35: © Gerald A. DeBoer – Shutterstock

Table of Contents

INTRODUCTION

It's pretty easy for people to cope with winter's cold. Feeling chilly? Put on a sweater. Have to go outside in the snow? Bundle up in a coat, scarf, hat, and boots. And when you come back inside, you can enjoy your heated home. You might even treat yourself to a bowl of hot soup or a mug of hot chocolate.

Animals have a harder time staying warm. Sure, pet cats and dogs can curl up inside. But what about bears and frogs, birds and bees? These animals are stuck outside. They have to find shelter, store food, and keep warm. Cold-blooded animals, such as reptiles and amphibians, have the biggest challenge. How do you stay warm when the air around you is freezing?

Yellow-bellied marmot

Animals have come up with all sorts of weird and wonderful ways to get through the winter. Some sleep. Others create body heat in unusual ways. Some animals even freeze almost solid, then thaw out when warm weather wakes them up.

So sit back, warm up, and discover all the amazing ways animals stay alive no matter what winter has in store.

Chapter One

BODY HEAT CAN'T BE BEAT

The wind is blowing, The temperature is dropping. What's an animal to do?

Size Matters

Many mammals that live in cold places are bigger than their relatives that live in warm places. For example, the Arctic-dwelling polar bear is much larger than the sun bear, which lives in the hot rain forests of southeast Asia. Why is size important? The answer is that more of a large animal's body comes in contact with the cold. This spreads out the cold air, so the animal can distribute its body heat and stay warm.

Polar bear

Small body parts are more likely to freeze or be damaged by the cold. To avoid this, many cold-climate animals have small ears and tails. An animal called a pika looks like a small rabbit. Take a closer look and you'll see that a pika has tiny ears that look very different than a rabbit's long ears. The pika also has no tail. These features help the pika keep its body warm and not have to worry about freezing its tail—or ears—off!

American pika

Terrific Tails

Many mammals use their own bodies to stay warm. Look at the red fox's big, fluffy tail. When it's cold, the fox wraps that tail over it's face, just like you might cover your nose and mouth with a scarf. The thick fur blocks the wind. It also keeps the fox's body heat next to its head instead of escaping into the cold air.

Red fox

Snow leopards are also part of the Big Tail Club. These big cats live high in the Himalayan Mountains, where it is really, really cold. When it's freezing and snowing, the cat covers its face with its tail to block the wind and the cold.

Another wild cat uses its tail in a different way. The Pallas' cat lives in many cold areas in western Asia. To keep its feet off the cold ground, it stands on its thick, plush tail.

Snow leopard

Pallas' cat

How To Keep Feet From Freezing

An animal's feet can be one of the most exposed parts of its body. As the animal walks around, its feet or paws get quite a chill. If paws or toes get too cold, they can develop **frostbite**. That's when the animal's skin or body part actually freezes. That's very painful—and very dangerous! If an animal loses part of its feet, it won't be able to walk.

Here's the good news: Some animal bodies have come up with clever ways to avoid frostbite. Take a look at the Arctic fox. This animal lives in the frozen north. It does just fine in temperatures as low as -58° F (-50° C). How can the fox walk around in such cold weather? The thick fur on its paws help the fox keep warm. There's even fur between its toes! Even better—its body **circulates** blood away from its feet. How does that help? It lowers the temperature of the fox's feet. That means the skin and toes are less likely to freeze.

Arctic fox

Penguins are another animal that reduce the blood flow to their feet. By sending less blood to their feet, a penguin's body can hold onto heat longer.

Some species of flamingoes and ducks do just the opposite to keep their feet warm. These birds don't live where it's super cold, like Arctic foxes and penguins do. So it's not likely their feet will freeze. Still, it's no fun to stand around on cold toes. So these birds increase the flow of blood to their feet. The blood warms up the feet as it circulates and keeps the birds toasty and comfy.

FUN FACT

A snow leopard's paws are wide and furry. Their feet are like snowshoes. This helps them walk on top of thick snow without sinking.

Flamingoes

Snow leopard

Fur Coats and Feathers

Many animals grow thicker fur during the winter. Just like putting on a warm coat, that thick fur holds in body heat and keeps the animal warm. Many mammals use this trick, including squirrels, bison, wolves, wild cats, and bears.

Mountain goats have two layers of fur. The top layer of fur is warm and wooly. The undercoat is made of hollow hairs that trap body heat. Heat stays close to the body, while cold and snow stay out. Moose also have this type of double coat.

Mountain goat

What about birds? These animals don't have fur, but they still need to keep warm. In this case, feathers do the trick. Have you ever seen a bird all puffed up on a cold day? Birds fluff out their feathers to keep out the cold. Under the feathers, body heat stays where it belongs— close to the bird's body.

Robin

Group Hug!

When you can't make your own body heat, you have to rely on your friends. That's what garter snakes do. Snakes are cold-blooded, which means they can't control their body temperature. That's a big problem when it's below freezing outside.

Garter snakes **hibernate** in snug places, like an animal **burrow** or a space under some rocks. To stay warm, groups of snakes curl up together in a big ball. They share their body heat as they sleep through the winter. A hibernating group can include hundreds or even thousands of snakes.

Some reptiles snuggle up with friends from other species. Scientists have found Eastern diamondback rattlesnakes hibernating in the same dens as gopher tortoises.

Eastern garter snakes

Emperor penguins also stay warm by staying close. These big birds stand on the ice all day, every day. They huddle up in a group to share their body heat. Emperor penguins also lean back on their tails to conserve heat and keep their feet off the ice.

Staying warm using body heat works great for some animals. But for others, like fish and frogs, that just doesn't work. These animals have come up with a whole different plan that's hard to believe. Turn the page to find out more.

FUN FACT

Shivering when you're cold helps your body warm up. Bees have their own way of shivering to stay warm inside their hives. They vibrate their bodies but keep their wings still. Those vibrations create heat and keep the hive toasty-warm.

Honeybees

Emperor penguins

Chapter Two

FROZEN FROGS

If you are a cold-blooded animal that relies on warm temperatures to stay alive, winter can be a deadly problem. During warm seasons, amphibians like frogs can lay out in the sun to keep their body temperature nice and warm. The same is true of lizards, who can often be seen **basking** in the sun. What happens when winter comes and the temperature drops? Don't worry. Amphibians and reptiles have a very surprising way to survive the winter. They don't worry about staying warm. Instead, they freeze!

The Magic of Sugar

Sweet things taste good. But too much sugar is bad for people. Fortunately for the wood frog, too much sugar is exactly what this creature needs to survive the winter.

Collared lizard

As the weather gets cooler, wood frogs get hungry—REALLY hungry! They eat everything they can find. Their bodies get bigger. But the frogs aren't storing fat, like a hibernating bear does. They are storing a simple sugar called **glucose**.

When glucose is stored, it has a different name. It is called **glycogen**. Wood frogs fill their bodies with glycogen. When it gets cold, they find somewhere safe to stay, such as under a pile of leaves. Then something amazing happens. These frogs turn into little green ice cubes!

The water inside the frog's body freezes. As this happens, the cells shrink. But those cells are full of glycogen, plus a little bit of water. That water bonds, or sticks, to the glycogen instead of turning into ice. When spring comes, the frogs thaw out and hop away.

Wood frog

Beware: Ice Knives!

What happens when the water in blood freezes? It isn't pretty. The water turns into ice crystals. Ice crystals are very, very sharp. They can tear the body's cells like tiny knives or razors. The animals are stabbed to death from the inside out. Gruesome, isn't it?

Fortunately, ice crystals can't form with a lot of glucose stored in the cells. Like the wood frog, the European common lizard also stores a lot of glycogen in its cells. The glycogen prevents ice crystals from forming and keeps the lizard alive until it thaws out in the spring.

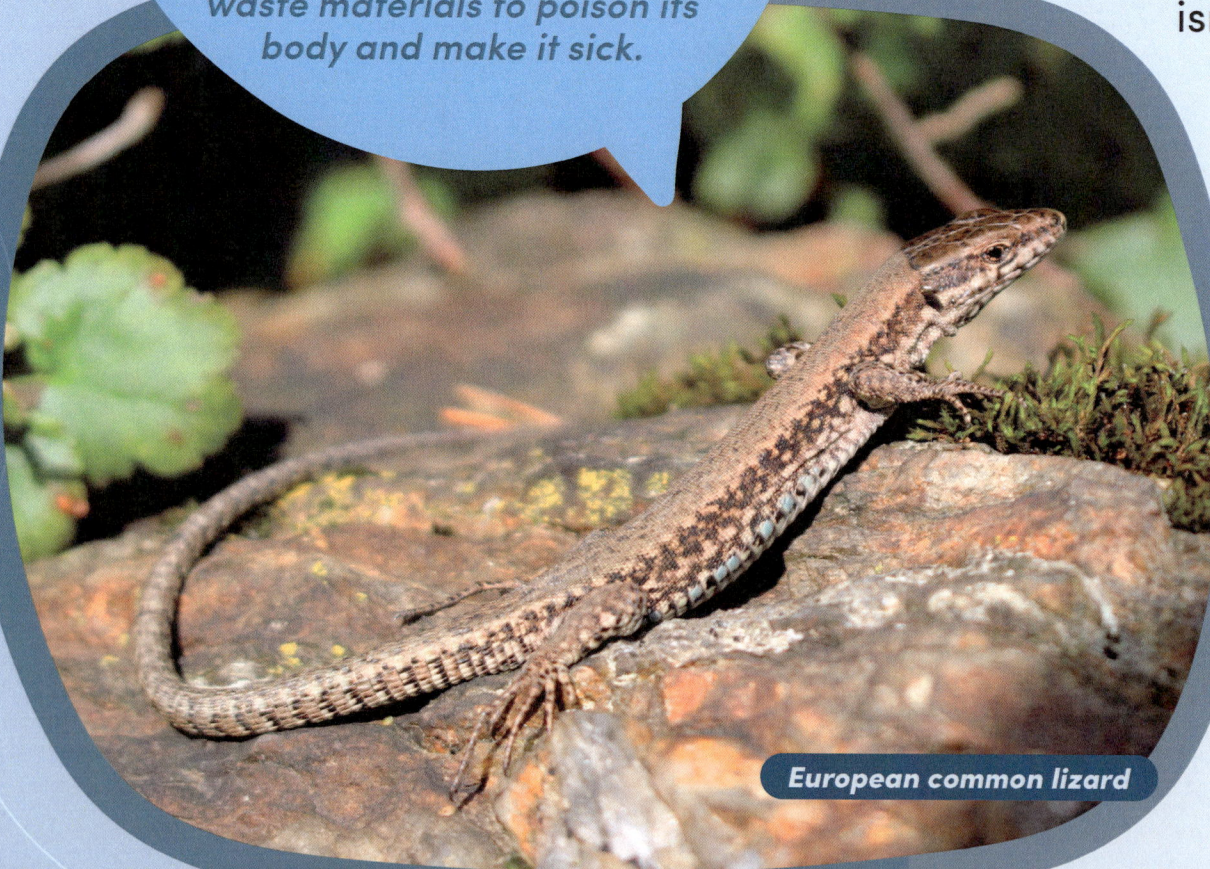

European common lizard

Slow Down!

Another way reptiles stay alive in the freezing cold is to slow down their body's activities. Many freshwater turtles can slow down their bodies' **metabolism**. Their hearts beat just a few times a minute. They take only a few breaths. Some of their organs shut down completely. These measures sound crazy, but they allow the turtle to use a lot less energy. Staying warm uses a lot of energy, so it's a big help for the turtle to reduce its energy needs.

Freshwater turtles

19

Going Deep

Some animals survive winter by digging deep. The frost line marks how deep soil freezes in the winter. Good diggers, such as the American toad and the spotted salamander, simply dig burrows deeper than the frost line. It is warm enough down there that these animals will not freeze to death.

Other animals go deep in the water. Lakes and ponds may have a layer of ice on the surface, but the water isn't frozen underneath. Bullfrogs and painted turtles simply hang out underneath the ice. Sometimes you might even see turtles swimming around down there.

FUN FACT

How can air-breathing animals survive underwater? Some **aquatic** *animals can also breathe through their skin. Their bodies take oxygen out of the water, just like a fish does with its gills.*

Bullfrog

Fish also live in water, which does, of course, freeze when the temperature drops. So what's a fish to do? The answer lies deep beneath the surface.

Here's something you might not know. The ocean, along with most lakes and rivers, does not freeze solid. A type of fish called the icefish lives in the waters around the South Pole. These fish dive deep beneath the surface. The water there is warm enough to let the fish survive.

Another fish, called the northern cod, lives on the other side of the world from the icefish. Northern cod live in the chilly waters of the Atlantic Ocean. However, northern cod also know the deep-diving trick. They spend the winter deep underwater, where the water is warm enough to keep them alive.

The frogs and turtles in this chapter have crazy ways to stay warm. Other animals have a different plan. They deal with winter by sleeping through it!

Cod fish

Chapter Three

HIBERNATION NATION

Have you ever woken up on a cold, cloudy day and just wanted to pull the covers over your head and stay in bed? Many animals have the same idea. Let's meet some animals who sleep through the winter—and use some weird and wonderful tricks to stay alive.

Raccoon

Eat Like Your Life Depends on It

You probably know that animals have to live off of stored energy while they hibernate. Because true hibernators (more on that later) do not wake up, they do not consume any food. Without food, their bodies can't produce the energy they need to stay warm and stay alive.

To solve this problem, many mammals eat like crazy in the weeks before winter's cold sets in. In fact, pretty much all they do is eat. Their bodies get big and round with stored fat. During the winter, that fat provides the energy these chonky critters need to stay alive.

Fat-tailed dwarf lemur

23

The Power of Pee

Black bears have another trick to staying alive during the winter, and it's pretty gross! These animals hibernate for many months. Like other mammals, they eat a lot to build up fat reserves. They also grow a thicker winter coat to stay warm. But these big bears have one more trick up their furry paws!

If a person slept for four months, they would not feel very well when they woke up. Their muscles would waste away. They would be too weak to walk or even stand. Why doesn't this happen to black bears? The secret is in their pee!

Black bear

Black bears do not **urinate** while they hibernate. Instead, they absorb the pee into their blood. From there, the liquid travels to the liver. The liver removes a chemical called **urea**. Urea is full of **nitrogen**. This nitrogen helps keep muscles strong. That means that when the bear wakes up in the spring, it is ready to go outside.

FUN FACT

The urea in a bear's urine also helps them heal. Scientists studied bears that had cuts on their bodies when they started to hibernate. When the bears woke up, the cuts were gone.

Black bear

A Warm Winter Nap

Bears are not the only mammal that hibernates. Bats do too. Large groups of bats find a warm, safe space to sleep. They might hibernate in a cave. Or they might find a cozy spot inside a building. The bats hang upside-down all winter long. In the spring, they wake up and fly away.

Small animals often go underground to stay warm while they hibernate. Moles live in underground tunnels. These tunnels are a great place to stay warm and store food for the long winter. Hiding underground also keeps these animals safe from predators.

Greater horseshoe bat

Other animals are not true hibernators. They do not sleep all winter. Squirrels and raccoons do fall asleep when it's very cold out. But on a warm day, these creatures wake up from their nap. They go outside and have a snack before hiding in their nests again. After all, wouldn't you be hungry after a long nap?

Hibernation is a great plan for some animals, but not all. Turn to the next chapter to find out some other clever ways animals have found to survive cold weather.

FUN FACT

Mice can live between the ground and the snow. The snow insulates the ground. This insulation holds in heat and creates a layer that is warm enough for the mice to stay warm.

Brown rat

Gray squirrel

Chapter Four

FINDING WARMTH

Of course, not all animals hibernate. Even growing a thick winter coat isn't always enough to survive the bitter cold. Never fear! Animals have come up with some clever ways to find the warmth they need to stay alive.

Soaking Up the Warmth

Some places on Earth have hot springs. These are areas of super-hot water that bubble up out of the ground. If an animal is lucky enough to live near a hot spring, you can bet it will take advantage of the warmth.

Yellowstone National Park is located in the western states of Wyoming, Montana, and Idaho. It is full of hot springs. The water in these springs is much too hot to touch. However, the water also warms the ground around it. The bison and elk that live in the park can often be seen lying on the warm ground near these hot springs.

Bison

Macaques are a type of monkey. They live all over Asia. Most places where they live are warm. However, macaques that live in northern Japan have to face the cold and snow. These monkeys have come up with a luxurious way to stay warm. They bathe in hot springs found in the mountains.

Soaking in hot springs appears to be a trick macaques teach one another. Scientists first saw macaques bathing in hot springs in 1965. They studied the monkeys for many years. These scientists discovered that mother macaques taught their babies to use the springs. Macaques that moved into the area from other places did not use the springs. They are missing out on a great way to stay warm!

Japanese macaque

Hiding Inside

People stay warm in heated homes, stores, and businesses. Sometimes animals do too! As the weather gets colder, insects and spiders seek out warm places to spend the winter. Many come inside homes and workplaces. So next time you spot an insect or spider in your house, be kind. Just like you, the creature needs a warm, safe place to stay.

Mice also sneak inside houses. Attics are a great place for these little creatures to spend the winter. And since a mouse can squeeze through even tiny holes, it's pretty easy for them to find a way into any warm building.

Mouse

A reptile called the tuatara has a sneaky way to stay warm. These animals live in New Zealand. During cold weather, they often hide in seabird nests at night. They "steal" heat from the birds. This raises their body temperature and helps the tuataras stay warm during the day. And the birds don't seem to mind!

FUN FACT

*Stealing heat from another animals is called **kleptothermy**.*

Tuatara

Get Out of Town!

When all else fails, sometimes the best way to survive the winter is to leave town. Some people have winter homes in warm places. Animals do too!

Many birds migrate. They fly south, east, or west in search of milder weather. How do they know when it's time to leave? Scientists think the shorter hours of daylight and falling temperatures alert the birds that it's time to go.

Canada geese

Birds aren't the only animal that migrate. Elk and mule deer in Glacier National Park in Montana live high in the mountains during the spring and summer. When the weather gets cold and snowy, they move down to lower elevations. Here the weather is warmer and food is easier to find.

Alpine swift

Mule deer

Insects migrate too. One of the most famous migrators is the monarch butterfly. They are the only butterfly that migrates to warmer climates and back home. Monarchs that live east of the Rocky Mountains fly thousands of miles to Mexico for the winter. Monarchs in the western part of the U.S. spend their winters in California. Monarchs huddle together to share body heat. Tens of thousands of these butterflies can gather on a single tree.

Migration might seem like a great idea. However, it is very hard work. It takes a lot of energy to travel so many miles. There is also the problem of finding food on the way. And when the animals get to their winter homes, they have to find shelter and food in a place that already has other animals living there. Still, for many creatures, migration is the answer to surviving winter.

Monarch butterfly

So next time you put on your winter coat, mittens, and hat and venture out into the cold, think of all the animals in the world that have to stay warm. Whether it's hibernation, migration, or changing their bodies, the creatures of the Earth know the best ways to survive the winter.

FUN FACT

Climate change has affected when and where animals migrate. In the northeastern part of the United States, robins and Canada geese used to fly south for the winter. Now, these areas have milder winters, so birds often stay in their northern homes all year long.

American robin

GLOSSARY

aquatic—living in water

basking—lying in the sun to stay warm

burrow—an underground home for an animal

circulates—moves through a system

frostbite—an injury to the body caused by extreme cold

glucose—a simple sugar

glycogen—stored glucose

hibernate—to spend the winter in a deep sleep

kleptothermy—stealing another animal's body heat to keep warm

metabolism—the processes that take place in a body to keep it alive

nitrogen—a gas present in the body

urea—a chemical compound found in urine

urinate—to get rid of liquid waste

ABOUT THE AUTHOR

Joanne Mattern has written more than 350 books for children. Many of her books focus on science, social studies, history, and biography, and her publishers include Running Press, Enslow, Lerner, and Scholastic.